The Earth Moves

PUBLISHED TITLES IN THE GREAT DISCOVERIES SERIES

DAVID FOSTER WALLACE
Everything and More: A Compact History of ∞

SHERWIN B. NULAND
The Doctors' Plague: Germs, Childbed Fever, and the Strange Story of Ignác Semmelweis

MICHIO KAKU
Einstein's Cosmos: How Albert Einstein's Vision Transformed Our Understanding of Space and Time

BARBARA GOLDSMITH
Obsessive Genius: The Inner World of Marie Curie

REBECCA GOLDSTEIN
Incompleteness: The Proof and Paradox of Kurt Gödel

MADISON SMARTT BELL
Lavoisier in the Year One: The Birth of a New Science in an Age of Revolution

GEORGE JOHNSON
Miss Leavitt's Stars: The Untold Story of the Woman Who Discovered How to Measure the Universe

DAVID LEAVITT
The Man Who Knew Too Much: Alan Turing and the Invention of the Computer

WILLIAM T. VOLLMANN
Uncentering the Earth: Copernicus and The Revolutions of the Heavenly Spheres

DAVID QUAMMEN
The Reluctant Mr. Darwin: An Intimate Portrait of Charles Darwin and the Making of His Theory of Evolution

RICHARD REEVES
A Force of Nature: The Frontier Genius of Ernest Rutherford

MICHAEL LEMONICK
The Georgian Star: How William and Caroline Herschel Revolutionized Our Understanding of the Cosmos

FORTHCOMING TITLE:

Lawrence Krauss on the science of Richard Feynman

GENERAL EDITORS:

Edwin Barber and Jesse Cohen

BY DAN HOFSTADTER

Falling Palace: A Romance of Naples

The Love Affair as a Work of Art

Goldberg's Angel: An Adventure in the Antiquities Trade

Temperaments: Artists Facing Their Work